SpringerBriefs in Biotech Patents

Series Editor

Ulrich Storz, Duesseldorf, Germany

For further volumes:
http://www.springer.com/series/10239

Andreas Hübel · Ulrich Storz
Aloys Hüttermann

Limits of Patentability

Plant Sciences, Stem Cells
and Nucleic Acids

Springer

Andreas Hübel
Patent Attorneys
  Michalski Hüttermann & Partner
Duesseldorf
Germany

Aloys Hüttermann
Patent Attorneys
  Michalski Hüttermann & Partner
Duesseldorf
Germany

Ulrich Storz
Patent Attorneys
  Michalski Hüttermann & Partner
Duesseldorf
Germany

ISSN 2192-9904   ISSN 2192-9912   (electronic)
ISBN 978-3-642-32840-4   ISBN 978-3-642-32841-1   (eBook)
DOI 10.1007/978-3-642-32841-1
Springer Heidelberg New York Dordrecht London

Library of Congress Control Number: 2012946185

The information provided herein reflect the personal views and considerations of the authors. They do not represent legal counsel and should not be attributed to the companies or law firms the authors work for.

Printed on acid-free paper

Springer is part of Springer Science+Business Media (www.springer.com)

Preface

This is the third volume of SpringerBriefs in Biotech Patents. The overarching topic in which the three articles comprised herein have in common relates to the Limits of Patentability—a topic which is often raised when it comes to Biotech patents.

Until recently, newly created novel embodiments in Biology were excluded from patentability, as the classical breeding methods used therefore relied on the random distribution of genetic matter, and thus did lack reproducibility—which is seen as one condition required to confirm technicity, which again is, at least in Europe, a *conditio sine qua non* for patentability.

With the rise of biotechnological methods, such as restriction enzymes, PCR, transfection methods, and the like, a molecular toolbox is now available which provides reproducibility with a sufficiently high degree. Patent applications related to these methods do therefore comprise a clear technical teaching—with the result that technicity is no longer denied for most biotechnological methods.

However, biopatents do also have to face challenges on other grounds. In many industrialized markets a strong public movement exists not only against biotechnology as such (with an emphasis against so-called "green" biotechnology), but in particular against patents for biotechnological inventions. As regards the latter case it is often overlooked that patents do not provide a right to practice of the protected invention, but only an exclusive right under which the patentee can block others from using said invention.

Some disciplines in biotechnology do, without doubt, raise new ethic questions on which most societies have no consensual answers yet. However, in their helplessness, societies tended to seek answers on these questions in the patent law. As a result, the number of special regulations which, for example, the European Patent Convention provides for biotechnology inventions exceeds those for any other discipline.

However, the mere exclusion of particular types of invention from patentability due to ethical concerns does not automatically result in a ban of these inventions from being practised. The public discussion around the exclusion of particular biotech inventions from patentability is thus a mock battle.

The present volume tries to give a state-of-the-art overview of patentability issues in plant biosciences, stem cells, and nucleic acids. The authors hope to create a better understanding of these currently debated issues, and to help the reader to objectify his opinion on questions of patentability in these technical disciplines.

Duesseldorf, 5 July 2012

Ulrich Storz
Andreas Hübel
Aloys Hüttermann

Contents

The Limits of Patentability: Plant Biosciences

Andreas Hübel

Abstract Current issues in the patentability of plants produced by essentially biological processes within the EP and the controversy between farmers' privilege and patent exhaustion with respect to seeds in the US are discussed in this chapter.

Keywords Essentially biological processes · Product-by-process claims · Farmers' privilege · Patent exhaustion

1 Introduction

Agriculture, gardening, and floristics are significant markets having for example an annual turnovers of more than 1 billion US$ with seeds[1] or a market volume of turnover of about 9.1 billion Euro for alive plants in Germany only.[2] To meet the market needs for ever new and improved varieties which are more resistant against environmental stresses, give higher yields, possess better processability in industrial manufacturing of food, include an optimized content of specific metabolites, have more appealing blossoms, and many more, the plant breeding industry is highly innovative, investing about 16 % of its earnings in science and developing new plant varieties.[3] Today's plant breeding comprises cutting edge technology,

---

[1] http://www.bdp-online.de/en/Branche/Kennzahlen/
[2] Branchenfocus Lebendes Grün 2010; Verlag BBE medi; Neuwied, Germany.
[3] http://www.bdp-online.de/en/Branche/Kennzahlen/

A. Hübel (✉)
Michalski Hüttermann and Partner Patent Attorneys, Neuer Zollhof 2,
40221, Düsseldorf, Germany
e-mail: huebel@mhpatent.de

A. Hübel et al., *Limits of Patentability*, SpringerBriefs in Biotech Patents,
DOI: 10.1007/978-3-642-32841-1_1, © The Author(s) 2013

and despite the time it usually takes from the beginning until a new variety enters the market—about 25 years—plant breeding is a highly competitive field.

There is no doubt that the effort of any breeding company in providing a new variety should be protected such that said company can receive a return of its investment by adequate commercialization of the new variety. Intellectual property rights provide a basis for securing a refund of the expenditures that are necessary for making inventions or developing new plant varieties. The primary intellectual property right for protecting new plant varieties appears to be the "plant variety protection" which is intended to protect the unique genetic combination of a new variety and the resulting properties of the plant. Plant variety protection is available in a number of countries, such as China, India, Indonesia, Thailand, Brazil, Australia, USA, Israel, and European countries. In fact, 70 states are members of the International Union for the Protection of New Varieties of Plants (upov) [4]

In principle, plant variety protection will only be granted if the new variety is novel, distinct, uniform, and stable. In addition, the new variety must have a plant variety denomination. Plant variety protection is usually granted for 25 years, and protects the production, marketing, and import or export of propagation material of the protected variety.

A peculiarity of plant variety protection is the "breeders' exemption". Thereby, breeders are allowed to use a protected variety of their competitors for developing new varieties without being sanctionable by the holder of the plant variety protection right. This measure shall secure that plant breeders have access to the most up-to-date genetic material such that they can constantly offer a large variety of new varieties.

Nowadays, plant breeding utilizes biotechnological methods. This does not necessarily mean that the plants are genetically modified. Methods such as "marker-assisted breeding" speeds up the breeding process, because plants can be selected by their genetics much earlier than by their phenotype. Briefly, plant breeders make inventions and develop new breeding techniques which are not restricted to a particular plant variety. Plant variety protection is not applicable in such cases, but patent protection may be possible. The transposition of the Bio-patent Directive of the European Union into regional and national patent law provides an effective legislative tool. For example, it provides a farmer's privilege such that a farmer can keep seeds of patent-protected plants for subsequent plantings. The transition also extended research exemptions so as to allow that patent-protected plants are used for plant breeding without the need for prior consent of the patent holder.

The patent offices take care of the patent application's examination to ensure that the patentability criteria as novelty, inventiveness, and industrial applicability have been fulfilled for inventions concerning plants and methods for producing

---

[4] http://www.upov.org/export/sites/upov/members/en/pdf/pub423.pdf

plants. However, biotechnological inventions and inventions in the field of plant biosciences are confronted with some peculiarities in patent laws. Some of these peculiarities are discussed herein below.

2 Under EPC Plants are Patentable but Plant Varieties as Such are Not

The European Patent Convention provides in Article 53 (b) that European patents shall not be granted in respect of plant varieties or essentially biological processes for the production of plants. The same Article further provides that this exception of patentability shall not apply to microbiological processes or the products thereof.

In its decision G 1/98,[5] the Enlarged Board of Appeals (EBA) at the European Patent Office, which clarifies and interprets important points of law relating to the EPC, and ensures uniform application of the law, ruled that a claim wherein specific plant varieties are not individually claimed is not excluded from patentability under Article 53(b) EPC even though it may embrace plant varieties. The exception to patentability in Article 53(b), first half-sentence, EPC applies to plant varieties irrespective of the way in which they were produced. Therefore, plant varieties containing genes introduced into an ancestral plant by recombinant gene technology are excluded from patentability.

Thus, the Enlarged Board of Appeals had decided that no European patent shall be granted for claims directed to a specific plant variety. However, if the claims are directed to a higher order within the system, thus being directed at least to a species of plants, those claims may be granted. It does not matter, whether the plants were obtained by conventional breeding or are transgenic plants. Hence, no European patent shall be granted for transgenic plant varieties, but not because of being transgenic, but for being a plant variety.

Moreover, the EBA ruled that Article 64(2) EPC is not to be taken into consideration when a claim to a process for the production of a plant variety is examined. Article 64(2) EPC provides that the protection conferred by a European patent shall extend to the products directly obtained by the process, if said process is subject matter of the European patent. Given that Article 53(b) EPC excludes plant varieties from being patented, but no legal provision within the EPC provides that European patents shall not be granted for processes for the production of a plant variety. Hence, taking Article 64(2) EPC into consideration, one might assume that the prohibition of European patents for plant varieties can be circumvented by claiming the process for producing said plant variety—at least if said process is not an essentially biological process (see herein below). However, the EBA clarified that this shall not be possible.

---

[5] ABl. EPA 2000: 111.

3 Breeding Methods are Not Patentable Under EPC

Another exception of patentability within the EPC provides that no European patent shall be granted for essentially biological processes (Article 53(b) EPC). However, until recent decisions G 2/07[6] (Broccoli case) and G1/08[7] (Tomato case) of the EBA, it was not clear what in fact an "essentially biological process" is. Does "essentially" means exclusively or solely such that any technical and hence non-biological step within the process would render that process being not essentially biological. That would mean that introducing a single technical step into a process comprising otherwise only biological process steps would overcome the patentability exception, just as it is. Or should "essentially" be understood as pertaining to the gist of the invention? The latter understanding would pose tremendous problems in interpreting the invention and would require valuing the invention too.

In late 2010, the Enlarged Board of Appeal (EBA) of the European Patent Office released its decisions G 2/07 and G 1/08 with respect to "essentially biological processes for the production of plants" elucidating what kind of processes for producing plants are excluded from patentability under Article 53(b) EPC as being essentially biological processes, and attempts to clarify the meaning of the expression "essentially biologically".

The question whether a non-microbiological process for the production of plants which contains or consists of the steps of sexually crossing the whole genomes of plants and of subsequently selecting plants is in principle excluded from patentability as being "essentially biological" within the meaning of Article 53(b) EPC was answered in confirmative way. The Enlarged Board of Appeal clarified that in principle any non-microbiological process for the production of plants comprising the steps of crossing and selection is excluded from patentability under Article 53(b) EPC. Moreover, such a process does not escape the exclusion of Article 53(b) EPC only because it contains, as a further step or as a part of any of the steps of crossing and selection, a step of a technical nature which serves to enable or assist the performance of the steps of sexually crossing the whole genomes of plants or of subsequently selecting plants. Beyond that, for excluding such a process from patentability as being "essentially biological" within the meaning of Article 53(b) EPC, it is not relevant whether a step of a technical nature is a new or known measure, whether it is trivial or a fundamental alteration of a known process, whether it does or could occur in nature or whether the essence of the invention lies in it.

Hence, it does not matter whether the claimed process solely consists of the steps of sexually crossing whole genomes of plants and subsequently selecting

---

[6] http://documents.epo.org/projects/babylon/eponet.nsf/0/791D677646A4A968C12577F4004C3 445/$File/G2_07_en.pdf

[7] http://documents.epo.org/projects/babylon/eponet.nsf/0/E72204692CFE1DC3C12577F4004B EA42/$File/G1_08_en.pdf

plants having desired features, or whether the process comprises further steps of technical nature. Human intervention does not alter the principle that processes involving sexual crossing and selecting plants is an essentially biological process excluded from patentability under Article 53(b) EPC, even if said intervention was necessary for enabling the sexual crossing, if no pollination would occur without the human intervention or if the presence of a particular trait cannot be selected in any other way.

Thus, marker-assisted breeding is excluded from patentability under Article 53(b) EPC although the markers themselves constitute patentable subject matter, and may further constitute the gist of the breeding process. This exclusion has limited consequences, because only very few applications disclose markers. In the majority of cases, plant breeders prefer to keep their markers as secrets and are reluctant toward disclosing the markers they use to their competitors.

If, however, such a process contains within the steps of sexually crossing and selecting an additional step of a technical nature, which step by itself introduces a trait into the genome or modifies a trait into the genome or modifies a trait in the genome of the plant produced, so that the introduction or modification of said trait is not the result of the mixing of the genes of the plants chosen for sexual crossing, then the process is not excluded from patentability under Article 53(b) EPC.

In stating this way, the Enlarged Board of Appeal clarified that a breeding process overcomes its exclusion from patentability only if a trait is introduced or altered in the genome of the plant is when said plant is produced by means of a technical step within the steps of sexual crossing and subsequent selection such that the trait is not merely the result of the mixing of genes of the plants chosen for sexual crossing. Consequently, using a genetically modified parental plant in the crossing does not overcome the exclusion from patentability even if the process of the genetic modification is part of the claimed process. In addition, wherein the technical step of altering or introducing a new trait into the genome of a plant is performed prior to the sexual crossing and selection of the plants, the process comprising the technical step of altering or introducing a trait, sexual crossing of the genetically altered plant with another plant, and subsequently selecting for offsprings having the new or altered trait is not patentable. Such a process would violate the provisions of Article 53(b) EPC because introducing or altering the trait did not occur within the steps of crossing and selection.

4 Patentability of Product-by-Process Claims for Plants

Despite the clarification of the EBA concerning the exclusion of "essentially biological processes" from patentability, the controversy is continuing as the EPO allows patent applications for grant wherein said applications do not claim an essentially biological process for the production of plants, but a plant obtained by an essentially biological process.

A product-by-process claim defines a product by means of a process for its manufacturing. Such product-by-process claims are only allowable if the product is patentable as such, and if the product cannot be defined in a sufficient manner on its own, i.e., with reference to its composition, structure or other testable parameters. A product-by-process claim is in fact a product claim and is intended to provide the same absolute protection as a product claim does, i.e., without the limitation to the process specified in the claims, regardless of whether the open wording "obtainable by" or the closed language "obtained by" is used.[8] The provisions of Article 64 (2) EPC are not applicable in these cases.

Based on aforementioned provision, the principle that exceptions shall be interpreted strictly, granting claims directed to plants that are obtainable by an essentially biological process appears appropriate. On the other hand, opponents of this concept argue that the EPC provides in Article 64(2) EPC that if the subject matter of the European patent is a process, the protection conferred by the patent shall extend to the products directly obtained by such process, and that the exclusion of patentability for essentially biological processes also excludes patentability of the plants that can be obtained by such a process. The plants—at least the plant varieties—may be protected by the plant variety protection right. Thus, allowing claims for grant that are directed to plants characterized by an essentially biological process for their production would in fact bypass the exclusion from patentability, in particular if one considers that case law all over the world restricts the scope of protection conferred be a product-by-process claim to the process specified in the claim, independently of using open or closed language.[9]

For providing an answer, the question whether the exclusion of essentially biological processes for the production of plants in Article 53(b) EPC can have a negative effect on the allowability of a product claim directed to plants or plant material such as a fruit is referred to the Enlarged Board of Appeal.[10] More particularly, the EBA was asked whether a claim directed to plants or plant material other than a plant variety is allowable even if the only method available at the filing date for generating the claimed subject matter is an essentially biological process for the production of plants disclosed in the patent application. In addition, the EBA is requested to declare if it is of relevance that the protection conferred by the product claim encompass the generation of the claimed product by means of an essentially biological process for the production of plants excluded as such under Article 53(b) EPC.

It will be interesting to see whether product protection for plants other than plant varieties as such will remain for those cases wherein the plant cannot be

---

[8] Guidelines for Examination Part F-Chapter IV-15, 4.12.

[9] Polyäthylenfilamente, BGH X ZB 8/95, Kirin-Amgen, Inc vs. Hoechst Marion Roussell Limited, [2004] UKHL 46, Abbott Laboratories vs. Sandoz, Inc. [Fed. Cir. 2009], TEVA Gyogyszergyar vs. Kyowa Hakko Kirin [IP High Court of Japan 2010(Ne)10043].

[10] G 2/12.

characterized by any other mean than by a process of production when said process is an essentially biological process.

5 Farmers' Privilege and Patent Exhaustion

For generations, formers kept a portion of their harvested seeds for subsequent plantings. The patent law of many countries considers this habit in that they provide the so-called farmers' privilege which allows a farmer to also keep a portion of the harvested seeds which were obtained from plants that are protected by a patent. In return, the farmer compensates the patentee. However, once a product being protected by a patent has been sold or marketed by the patentee or under the patentees' approval, the patentee's patent rights exhausted such that the patentee cannot claim his patent rights for that particular product.

For any farmer, the question arises whether the seeds harvested from the crops that are (i) covered by the breeders patent rights and (ii) purchased by the farmer from the breeder or his licensee fall under patent exhaustion or not. In this regard, the US Supreme Court recently requested an opinion of the Solicitor General The Solicitor General's view is eagerly expected as it may provide an insight in how the Supreme Court might decide in the *Bowman versus Monsanto* case and more generally speaking whether self-replicating technologies are vulnerable to patent exhaustion or not.

Monsanto Co. sued the farmer Mr. Bowman for infringing Monsanto's patent rights on their Roundup Ready® technology,[11] because Mr. Bowman, who purchased Roundup Ready® soybean seed from Pioneer Hi-Breed, a licensee of Monsanto, from 1999 to 2007, planted the soybean seed under the license agreement required by Pioneer Hi-Breed as his 1-year planting. In addition, Mr. Bowman purchased commodity soybean seed from a grain elevator for a 2-year planting. As 94 % of soybeans sold into commodity markets in Indiana in 2007 used Monsanto's Roundup Ready® technology, it was no surprise that he found the commodity seeds showing the same herbicide resistance to glyphosate as the Roundup Ready® soybean seeds. Mr. Bowman then began saving part of his commodity seed harvest for subsequent plantings.

Mr. Bowman was honest with Monsanto about his use of the commodity soybean seed, but Monsanto investigated and—again not surprisingly—found out that the commodity soybean seed contained Monsanto's patented Roundup Ready® technology. Monsanto then sued Mr. Bowman for infringing above-identified patent rights. The district court ruled in favor of Monsanto and awarded damages.

In his appeal at the Federal Circuit,[12] Mr. Bowman argued that the doctrine of patent exhaustion applies to the authorized sale of seeds into commodity markets

---

[11] U.S. Patent 5,352,605 and U.S. Patent RE 39,247E.
[12] Monsanto Co. v. Bowman (Fed. Cir. 2011).

and any downstream product of purchases from these markets which possess essentially the same characteristics as the sale to the commodity market, i.e., the glyphosate resistance in the instant case. Monsanto on the other hand argued that patent protection is independently applicable to each generation of soybeans containing the patented trait.

The Federal Circuit found that the doctrine of patent exhaustion does not apply to the next generation of seeds, even if Monsanto's patent rights in the commodity seeds are exhausted. By planting the commodity seeds, a next generation of seeds is developed which contain Monsanto's Roundup Ready® technology. Thereby, a newly infringing article was created to which the doctrine of patent exhaustion does not apply. Hence, the Federal Circuit denied Mr. Bowman's view that each seed sold is a substantial embodiment of all later generations.

In its reasoning, the Federal Circuit considered that commodity seeds can be used in various ways, for example as feed. Using the commodity seed as feed or any other conceivable use wherein no replication of Monsanto's technology occurs would be free of liability for patent infringement. However, farmers can not replicate seeds including patented technology by planting them in the ground without creating newly infringing seeds.

In late October 2010, Mr. Bowman filed a petition for a writ of certiorari with the US Supreme Court which has taken the court's interest as can be inferred from request the Solicitor General's opinion. If the Supreme Court will deny certiorari, protection for transgenic plants is not just strengthened, but the rights of patentees in the field of biotechnology are strengthened, because the considerations apply to all replicative technologies. However, if the Supreme Court will reverse the Federal Circuit on the question of whether the doctrine of patent exhaustion is applicable to self-replicating technologies, any consumer and any competitor is able to avoid patent infringement by simply duplicating patented technology by means of growing or cultivating a sample purchased in the stream of commerce. A decision in the latter sense would devastatingly affect the position of any plant breeder as the intention of intellectual property rights would by annulled.

6 Conclusion

In plant bioscience, the breeders and inventors have to cope with peculiarities in patent protection. They may have an additional intellectual property right, the plant variety protection right, which mainly covers phenotypically distinct new varieties, but also need appropriate patent protection. In the near future, case law will provide some certainty to the breeders as to what is patentable and what scope of protection is conferred by the patents such that tailor-made strategies for protecting the business investments can be developed.

The Limits of Patentability: Stem Cells

Ulrich Storz

Abstract This chapter discusses the status quo in the patentability of human embryonic stem cells in Europe and the United States. Further, alternative technologies will be considered with respect to practicability and patentability.

Keywords Stem cells · Patentability · Embryonic · WARF · Brüstle

1 Introduction

According to recent reports, the market for stem cell technologies will grow quickly within the near future. Although projected figures are subjected to significant variances (business information provider Visiongain Ltd (2012) predicts that the overall world market for stem cell technologies in medicine will reach $7.3 billion in 2014, while competitor Kalorma Information is more conservative, yet its estimate still predicts that the global market for stem cell technologies will rise over $700 million in 2012, and given some positive trends could reach over 1 billion dollars in the same year.

Practising stem cell related technologies, particularly the derivation thereof, is subject to a strong regulation in most developed countries. Large differences apply from country to country, e.g., whether the importation of hES cell lines or their derivation within a country is legal. To make it more complicated, some countries have even intranational differences. In the United States, for example, particular

U. Storz (✉)
Michalski Huettermann and Partner Patent Attorneys, Neuer Zollhof 2,
40221, Duesseldorf, Germany
e-mail: st@mhpatent.de

A. Hübel et al., *Limits of Patentability*, SpringerBriefs in Biotech Patents,
DOI: 10.1007/978-3-642-32841-1_2, © The Author(s) 2013

aspects of practising stem cell related methods are subject to state law, with large differences from state to state (Caulfield et al. 2009).

In any case, stem cell related technologies comes at the expense of tremendous costs, which makes meaningful patent protection an important condition for investors to decide whether they may want to spend money into stem cell R&D.

However, the patentability of stem cells, particularly of human embryonic stem cells (hES cells), is an issue which has been discussed both in the public as well as in the biopatent community in all major industrialized countries. Key aspects of the discussion circle around ethical concerns related to the respective technologies and to the monopolization and commercial use of such cells, and the therapeutical promises made by these approaches. The following article will give an overview of the actual state of the patent debate, and the recent case law, with respect to Europe, and the United States.

2 The Legal Framework for Patentability of hES Cells

Stem cell related technologies do, without doubt, raise new ethic questions on which most societies have no consensual answers yet. However, in their helplessness, societies tend to seek answers on these questions in the Patent Law. As a result, the number of special regulations which, for example, the European Patent Convention provides for biotechnology inventions exceeds those for any other technical discipline.

In all discussions related to ethical issues of stem cell patents it should, however, be kept in mind that a patent is not a right to practice, but an exclusive right, i.e., a right to exclude others from practising an invention. The right to practice is dependent on (a) the respective legislation[1] and (b) existing patents of third parties. In case, a society may want to ban particular types of inventions which are deemed to be ethically problematic from being put into practice, the exclusion thereof from patent protection is, thus, an unsuitable tool.

2.1 Europe

As set forth in the European patent convention (EPC) and in the German Patent Act (PatG), patents are being granted for inventions, which are novel, rely on an inventive step and are industrially applicable. Both legal frameworks comprise generic clauses according to which inventions the commercial exploitation of which would be contrary to ordre public or morality are exempt from patent protection (Art 53 a) EPC, §2 (1) PatG). However, such exploitation shall not be

---

[1] In Germany, the derivation and use of hES cells is regulated by the German Stem Cell Act, while the use of embryos is regulated by the German Act on the Protection of Embryos.

deemed to fall under this exemption only because it is prohibited by law or regulation in some or all of the Contracting States. Accordingly, the Courts in the EU member states and the Technical Boards of Appeal of the European patent office (EPO) have rarely made use of said general clause.

In all cases, the question whether a given technology falls under this exemption requires a careful weighing up of the invention's usefulness to mankind against severity of the violation of *ordre public*. Cases where a given technology was considered to fall under this exemption encompass a coffin which could be evacuated to exclude that a seemingly dead wakes up after being buried,[2] and some biotechnological inventions in particular, e.g., a transgenic animal having increased probability of developing cancer.[3] Such mammal was claimed to be useful for cancer research, but since it could not be assumed that the only examples in the application, namely mice, could be extended to other animals, the Board of Appeal required that the claims are restricted to mice, because the patent would otherwise protect methods applied to animals other than mice (e.g., beavers), where the suffering involved would not be justified by sufficient benefit for mankind.

In Europe, biotechnological inventions are subject to European Directive 98/44/EG ("Biopatent Directive") since July 1998. The directive has subsequently been implemented into the respective laws of the EU member states as *lex specialis* over the generic exclusions as to "*ordre public*" discussed above.

Furthermore, the EPO has also implemented those clauses provided in the Regulation which refer to questions of patentability, although the EPO is not a body of the European Union, and was thus not obliged to do so. Interestingly, this means that regulations issued by the European Union became applicable law to non-EU states, as for example Switzerland, Turkey of Norway.

The key regulations set forth by the Biopatent Directive with respect to stem cells are as follows:

2.1.1 Positive Definition

A a positive definition, Art. 5 (2) sets forth that an element isolated from the human body may constitute a patentable invention, even if the structure of that element is identical to that of a natural element. This provision is commonly seen as the basis for the patentability of cells as such, including human cells.

2.1.2 Exclusions from Patentability

However, according to Art. 5 (1), the human body at the various stages of its formation and development cannot constitute patentable inventions. According to

---

[2] Decision of the German Patent Court, BPatG 23 W (pat) 248/70.
[3] EPO technical Board decision T 0315/03 (Transgenic animals/HARVARD).

Table 1 Overview of stem cell related clauses in the Biopatent Directive 98/44/EG

Directive 98/44/EG	Legal text	Implementation into the EPC
Art 5(1)	The human body at the various stages of its formation and development [...] cannot constitute patentable inventions	Rule 29 (1)
Art 5(2)	An element isolated from the human body [...] may constitute a patentable invention, even if the structure of that element is identical to that of a natural element	Rule 29 (2)
Art 6	Inventions shall be considered unpatentable where their commercial exploitation would be inconsistent to public policy or morality. The following, in particular, shall be considered unpatentable:	Art 53 a
	(a) methods for cloning human beings	Rule 28 (a)
	(c) the use of human embryos for industrial or commercial purposes	Rule 28 (c)

Art. 6, inventions shall be considered unpatentable where their commercial exploitation would be inconsistent to public policy or morality. In particular, Art. 6 sets forth that methods for cloning human beings (Art. 6(a)), and the use of human embryos for industrial or commercial purposes (Art. 6(c)) shall be considered unpatentable.

It is important to mention that, furthermore, Art. 6 sets forth that the said commercial exploitation shall not be deemed to be so contrary merely because it is prohibited by law or regulation. This means that particular indications supporting the *ordre public* issue are necessary to expel an invention from patentability. It is, however, common understanding that the examples mentioned in Art. 6 (particularly examples (a) and (c)) qualify as violating *ordre public* (see Table 1).

2.2 The United States

Compared with the extremely regulated situation in Europe, Title 35 of the United States Code has no specific exemptions for stem cell related patents. 35 U.S.C. §101 reads as follows:

> Whoever invents or discovers any new and useful process, machine, manufacture, or composition of matter, or any new and useful improvement thereof, may obtain a patent therefor, subject to the conditions and requirements of this title.

Obviously, 35 U.S.C. §101 mentions four categories of patentable subject matter, namely (any new and useful) process, machine, manufacture, or composition of matter. This list has long been interpreted as containing an implicit exception related to laws of nature, natural phenomena, and abstract ideas, which for a long time were deemed not patentable even by the US Supreme Court.[4]

---

[4] O'Reilly v. Morse, 56 U.S. 62 (1853).

Later on, the Courts recognized that too broad an interpretation of this exclusionary principle could eviscerate patent law. In *Diehr*[5] the Supreme Court pointed out that "a process is not unpatentable simply because it contains a law of nature or a mathematical algorithm, and that an application of a law of nature or mathematical formula to a known structure or process may well be deserving of patent protection." This increasingly liberal position culminated in Supreme Court decision *Diamond v. Chakrabarty*,[6] which issued in 1980, and according to which, as the Court put it, "Congress intended statutory subject matter to 'include anything under the sun that is made by man'".

Thus, no statutory exemptions from patentable subject matter exist for stem cell related inventions in the United States. Further, 35 U.S.C. §101 has no general exclusions as to "*ordre public*". Purified and isolated stem cells and human cloning-related inventions are thus considered patentable subject matter in the United States.

Recently, the Supreme Court has again restricted this very broad concept. In *Mayo Collaborative Services v. Prometheus Laboratories, Inc.*[7] a patent application related to a method of optimizing therapeutic efficacy for treatment of an immune-mediated disorder was deemed unpatentable because it was held that the claims effectively related to the underlying laws of nature themselves only. It is so far difficult to predict whether this decision will affect the patentability of hES cells. The decision is discussed in further detail below.

3 Case Law

3.1 Europe

The Biopatent Directive fails to properly define, among others, the term "embryo" and "human body". Some key issues related with the patenting of stem cell related inventions were thus unclear for a couple of years. These were, among others, the following:

(a) Does a given stem cell process involve the use of human embryos for industrial or commercial purposes ?
(b) Can a given stem cell as such be considered as being an embryo ?
(c) Can a given stem cell be considered as a human body at a stage of formation ?

These questions were addressed by the highest European authorities in Biopatent law, namely the Enlarged Board of Appeal (EBA) of the EPO and the European Court of Justice (ECJ). The two respective cases will be discussed in the following:

---

[5] Diamond v. Diehr, 450 U.S. 175 (1981).
[6] Diamond v. Chakrabarty, 447 U.S. 303, 308 (1980).
[7] Mayo Collaborative Servs. v. Prometheus Labs., Inc., 130 S. Ct. 3543 (2010).

3.1.1 The WARF Decision

In 2006, the EBA issued the so-called WARF decision.[8] The patent application in dispute (EP770125, inventor: James Thomson) was assigned to the Wisconsin Alumni Research Foundation (WARF). Its US counterpart has been nicknamed as "bottleneck patent" for all commercial stem cell products (Bergman and Graff 2007) due to its broad scope. The main claim of EP770125 was as follows:

A purified preparation of primate embryonic stem cells which

(a) is capable of proliferation in vitro culture for over 1 year,
(b) maintains a normal karyotype through prolonged culture,
(c) maintains the potential to differentiate to derivatives of endoderm, mesoderm, and ectoderm tissues throughout the culture, and
(d) will not differentiate when cultured on a fibroblast feeder layer.

The EBA has rejected the application due to a violation of Rule 28 (c) EPC, i.e., because it considered that the claimed embryonic stem cells involved, at least at the time of filing, the use of a blastocyst, which the EBA considered as an embryo. As the term "primate" encompasses "human", the EBA found that the criterion according to which a use of human embryos for industrial or commercial purposes is excluded from patentability was met.

The fact that the use of human embryos was not explicitly recited in the claims was deemed irrelevant, as the EBA considered the whole disclosure, not only the claims. Furthermore, the EBA stated that at the time of filing the cell cultures claimed could only be obtained by the blastocyst approach, which means that it required the destruction of human embryos.

The keynotes of the decision were as follows:

2. Rule 28(c) EPC forbids the patenting of claims directed to products which---as described in the application---at the filing date could be prepared exclusively by a method which necessarily involved the destruction of the human embryos from which the said products are derived, even if the said method is not pArt. of the claims.

4. it is not of relevance that after the filing date the same products could be obtained without having to recur to a method necessarily involving the destruction of human embryos.

The decision has often been interpreted as leaving room for patent applications related to the production of stem cells, or stem cells as such, if such application

---

[8] Decision G2/08.

describes at least one alternative way to produce the said cells (i.e., not related to, or involving, hES cells).

3.1.2 The Brüstle Case

Another case related to the patentability of hES cells has recently been decided by the ECJ. The case related to the German patent assigned to Professor Oliver Brüstle, who is a researcher at Bonn University, Germany. The patent was related to neural progenitor cells which have been derived from hES cells legally obtained under the deadline solution provided by the German Stem Cell Act (StZG).

In contrast to the WARF patent, this patent did, therefore, not relate to hES cells as such. While hES cells were disclosed as preferred source for the claimed neural progeny cells, cells obtained by parthenogenesis and cells obtained after somatic nuclear cell transfer were mentioned as alternative sources (although it seems to be arguable if they represent a technically feasible alternative).

Foreplay Under German Jurisprudence

The patent was granted in 1997 with the following main claim:

    Isolated, purified progenitor cells with neuronal or glial prop-
    erties of embryonic stem cells, comprising a maximum of 15 %
    primitive embryonal and non-neuronal cells, which are obtained by
    the following steps:

    (a) cultivation of ES-cells to obtain embryoid bodies,
    (b) cultivation of embryoid bodies to obtain neuronal progenitor
        cells
    (c) [...]

In 2004, Greenpeace filed a nullity suit against the patent, in the course of which the Federal Patent Court (BPatG) declared the patent invalid due to violation of § 2 (2) Nr. 3 PatG (which corresponds to Art. 6c Biopatent Directive) in 2006. The applicant went into appeal before the Federal Supreme Court (BGH).

Of course, the BGH has to apply German law and, as such, the EU Biopatent Directive when reexamining German patents. This means that in questions related to the patentability of German Biotech patents, the ECJ is the final instance for the interpretation of the respective rules.

For these reasons, the BGH decided on 12 November 2009 to stay proceedings and submit a referral to the ECJ for an interpretation of the Biopatent Directive, particularly of the terms "embryo" and "industrial"/"commercial".[9]

---

[9] BGH "Neurale Vorläuferzellen", Akz: Xa ZR 58/07.

In his referral, the BGH addressed different issues. First of all, the BGH wanted to know whether or not hES cells, cells obtained by somatic nuclear transfer, cells obtained by parthenogenesis and/or induced pluripotent cells (iPS) qualify as embryos (Art. 6) or as a human body at a stage of its formation and development (Art. 5).

Next, the BGH asked if cells which have been obtained directly or indirectly from hES cells are excluded from patentability (Art. 6) because, for the latter, an embryo was destroyed ("Fruits of the forbidden tree")?

The BGH, furthermore, noted that the research of Prof. Brüstle was publically funded and has, thus, been considered, at least once, to be in line with *ordre public*. It would thus be surprising, the BGH suggested, if same did not apply for resulting patent applications.

In addition, the BGH suggested, in his referral, that the term "embryo" should be defined according to § 8 of the German Act for the Protection of Embryos (EschG), which requires totipotency. This would mean that hES cells, which are not totipotent, do not qualify as embryos. However, the Act for the Protection of Embryos is a German law which is not an implementation of a European Directive, which means that the ECJ is not bound to said definition, and the BGH could only suggest to adopt the latter.

The Decision Issued by the European Court of Justice

In cases which raise a new point of law, decisions by the European Court of Justice are anticipated by an opinion issued by one Advocate General. Hence, as under the Biopatent directive, only one case had made it to the ECJ yet at that time,[10] which however was not related to hEScell issues, the Advocate General came again into play. On March 10, 2011, Advocate General M. Yves Bot recommended to answer the questions referred to the ECJ by the BGH with respect to Article 6(2)(c) of Directive 98/44/EC in that

- the concept of a human embryo applies from the fertilization stage to the initial totipotent cells and to the entire ensuing process of the development and formation of the human body, which includes the blastocyst;
- unfertilized ova into which a cell nucleus from a mature human cell has been transplanted or whose division and further development have been stimulated by parthenogenesis are also included in the concept of a human embryo in so far as the use of such techniques would result in totipotent cells being obtained;
- pluripotent embryonic stem cells are not included in that concept because they do not in themselves have the capacity to develop into a human being;
- an invention must be excluded from patentability where the application of the technical process for which the patent is filed necessitates the prior destruction

---

[10] Monsanto vs. Cefetra, C-428/08.

of human embryos or their use as base material, even if the description of that process does not contain any reference to the use of human embryos; and

• the exception to the non-patentability of uses of human embryos for industrial or commercial purposes concerns only inventions for therapeutic or diagnostic purposes which are applied to the human embryo and are useful to it.

The ECJ's decision issued 18 October 2011.[11] Not surprisingly, the ECJ essentially agreed with the Advocate General's opinion, and concluded that (i) any human oocyte after fertilization, (ii) a non-fertilized human oocyte into which a cell nucleus from a mature human cell has been transplanted, and (iii) any non-fertilized human oocyte the division and further development of which have been stimulated by parthenogenesis constitute a human embryo, and are thus excluded from patentability.

Further, the ECJ found that an invention is also excluded from patentability if the technical teaching requires prior destruction of a human embryo, or its use as base material, whatever the stage at which that destruction occurs, and even if said destruction is not part of the claimed technical teaching and does not refer to the use of human embryos.

The ECJ, furthermore, found that the exclusion from patentability concerning the use of human embryos for industrial or commercial purposes, as set out in Article 6(2) (c) of Directive 98/44/EC, also covers the use of human embryos for purposes of scientific research, because the grant of a patent already implies, in principle, its industrial or commercial application.

Still, patentability of inventions using human embryos is patentable for therapeutic or diagnostic purposes which are applied to the human embryo and are useful to it. However, it remains unclear what exactly is meant by the "use of a human embryo for therapeutic or diagnostic purposes which are applied to the human embryo and are useful to it".

With respect to pluripotent stem cells obtained from hES cells—which are not covered by the definition of "human embryo" as set forth by the ECJ because they do not qualify as totipotent—the ECJ ruled that it is for the referring Court to ascertain, in the light of scientific developments, whether such cells are capable of commencing the process of development of a human being and, therefore, are included within the concept of "human embryo" within the meaning and for the purposes of the application of Article 6(2)(c) of Directive of the Directive. Please note that these pluripotent cells are not to be confused with "induced pluripotent stem cells" (iPS cells), which are discussed herein below, and which have not been addressed by the ECJ.

---

[11] Oliver Brüstle vs. Greenpeace eV, C-34/10.

3.1.3 Differences Between Both Decisions

Interestingly, the ECJ's understanding of the term "human embryo" is consistent with that the EBA gave in the WARF decision. However, while the WARF decision has often been interpreted in such way that (i) patent applications which relate to inventions made after the underlying hES cell lines became available, and (ii) patent applications related to the production of stem cells, or stem cells as such, which describe at least one alternative way to produce the said cells (i.e., not related to, or involving, hES cells), are both patentable, such bypass is no longer possible in the understanding of the ECJ.

Although ECJ jurisdiction has no legal bearing for the granting practice of the EPO, the latter's president Benoît Battistelli announced in his Weblog,[12] on November 3, 2011, that "if the judges rule in favour of a restrictive interpretation of biotech patentability provisions, the EPO will immediately implement it". If any difference between the positions of the EPO and the ECJ has existed before, the EPO has thereby deliberately surrendered their position in favor of that of the ECJ.

As a result of the ECJ Decision, it is to be expected that the BGH will confirm the Germen Federal Court of Justice's declaration of invalidity of at least claims 1, 12, and 16 of German Patent No. 197 56 864.

3.1.4 Reactions by the Biotech Community

While the exemption from patentability does, as such, not affect the use of hES cells, it will affect the protection of research results, and thus may hamper the commercial exploitation of products and methods involving hES cells, and, hence, R&D related to hES cell-based therapies.

Not surprisingly, the Biotech community has reacted on the ECJ decision with intense indignation. In Germany alone, ten major research organizations, including the German Research Foundation (DFG), the University Rectors Conference and the Max Planck Society, published a joint statement in which they disapprove the decision, and polemize on its impact on stem cell research in Europe.

It will in fact be doubtable whether applied research on hES cells will still play a significant role in Europe. Sponsors will be hesitant to provide funds for applied research if it is already clear that the results of such research cannot be monopolized for a given period to recompensate the investments made. This means that applied research on hES cells will very likely decrease in Europe, thus turning Europe into a developing region at least with respect to this discipline.

Other voices state that the decision will not have any practical consequences, because it does not prohibit the respective stem cell related methods or products, but merely excludes the underlying inventions from patentability.

---

[12] see   http://blog.epo.org/uncategorized/patents-and-biotechnology-%E2%80%93-latest-developments/.

Because of the fact that in other major jurisdictions these limitations do not exist, so the argumentation (the US counterpart to the Brüstle patent, US 7,968,337, has been granted on June 28, 2011, i.e., about 4 months prior to the ECJ's decision and after the opinion of the Advocate General issued) a lack of protection in Europe alone would not be enough incentive for imitators to develop a counterfeit product exclusively for the European market. This amazingly calm position has been criticized as being mere calculated optimism in order to mollify investors, others say.

3.2 The United States

Like elsewhere, ethic questions are an evergreen issue in the history of stem cells in the United States, too. However, as set forth above, 35 U.S.C. §101 has no general exclusions as to *ordre public*, and no statutory exemptions from patentable subject matter exist for stem cell related inventions in the United States.

Only the Manual of Patent Examining Procedure (MPEP), which contains guidelines for the examiners at the USPTO, has codified, in section 2105, a clause according to which "if the broadest reasonable interpretation of the claimed invention as a whole encompasses a human being, then a rejection under 35 U.S.C. §101 must be made indicating that the claimed invention is directed to nonstatutory subject matter".

Further, the revision of the US Patent system in 2011 under the Leahy Smith Act, which came as quite a surprise, brought with it some significant changes. Among others, 35 U.S.C §101 will be amended by adding a passage according to which "no patent may issue on a claim directed to or encompassing a human organism."[13]

In an internal memorandum of September 20, 2011, the US Patent and Trademark Office (USPTO) informed its employees that this new clause merely codifies existing USPTO policy that human organism are not patent-eligible subject matter. It remains, however, to be seen, how the examiners put this clause into daily practice. Fears exist that stem cells, particularly hES cells, will sooner or later fall under this exemption as qualifying as a human organism.

3.2.1 Case Law with Respect to Patentability Issues

In two related cases concerning patent applications directed to a human/non-human chimera filed by Stuart Newman and Jeremy Rifkin, the USPTO rejected the both applications. In the first case[14] the USPTO emphasized that the

---

[13] Section 33(a) of the Leahy-Smith America Invents Act.
[14] US patent application No 08/993,563 filed by Stuart Newman and Jeremy Rifkin.

application would violate the utility requirement set forth in 35 U.S.C. §101, in that "inventions directed to human/non-human chimera could, under certain circumstances, not be patentable because, among other things, they would fail to meet the public policy and morality aspects of the utility requirement." The inventors refiled their application at a later stage, wit the same result. This time, the USPTO argued that although 35 U.S.C. §101 did not explicitly exclude the patentability of humans, the USPTO's position of rejecting such patent applications was implicitly encompassed by said statute.

The USPTO bypassed the Supreme Court's ruling in *Diamond v. Chakrabarty* by postulating that Congress could not have intended humans to be included as subject matter under 35 U.S.C. §101. This postulation resulted from an interpretation of the 13th Amendment of the US constitution, which abolishes slavery. Similarly, in *Tol-O-Matic Inc. v. Proma*[15] the USPTO stated that the utility requirement of 35 U.S.C. §101 excludes inventions deemed to be injurious to the well being, good policy, or good morals of society.

However, these decisions have not affected the general granting practice of the USPTO related to stem cells, which are granted by the USPTO on a regular basis (Bergman and Graff 2007). The real battles are today fought on other grounds.

The WARF Cases

One of the first applicants that had created a meaningful IP portfolio protecting methods for the production of hES cells is Wisconsin Alumni Research Foundation (WARF). Other players are Geron of Menlo Park, CA, and the NIH, who all stand in contractual relationships to one another, and have thus been nicknamed "gatekeepers of hES cell products" by some authors (Rabin 2005).

On July 17, 2006, Jeanne Loring, then associate professor at the Burnham Institute for Medical Research, Dan Ravicher, an attorney who had founded the Public Patent Foundation, and John Simpson of the Foundation for Taxpayer and Consumer Rights filed a request for re-examination of three WARF patents granted for methods related to the production of Primate Embryonic Stem Cells, namely US 5,843,780, US 6,200,806, and US 7,029,913, on the grounds of obviousness in view of published prior art.

On March 30, 2007, the USPTO rejected all three patents in their entirety on the grounds of obviousness. These decisions were appealed by WARF, with the result that US 5,843,780 and US 6,200,806 were upheld in amended form in March 2006, while the revocation of US 7,029,913 was confirmed on April 28, 2010, by the USPTO board of Appeals,

---

[15] Tol-O-Matic Inc. v. Proma Produkt-und Marketing Gesellschaft, 945 F.2d 1546 (Fed Cir 1991).

The Prometheus and BRCA Cases

Two other Court cases may have unprecedented effects on the patentability of stem cells, too.

In *Mayo Collaborative Services v. Prometheus Laboratories*,[16] the US Supreme Court overturned a prior decision by the Court of Appeals of the Federal Circuit, by judging that a method of optimizing therapeutic efficacy for treatment of an immune-mediated gastrointestinal disorder, as claimed in US Patent US6, 680, 302 assigned to Prometheus Laboratories, do not meet the patentable subject matter standard of 35 U.S.C. 101.

The claimed method comprised administering a given drug to a subject and determining its level, or of a metabolite thereof, in said patient, wherein a level below of a given threshold indicates to increase the dose of said drug and a level above the threshold indicates to decrease the dose thereof.

The Supreme Court considered that "the claims inform a relevant audience about certain laws of nature; any additional steps consist of well-understood, routine, conventional activity already engaged in by the scientific community; and those steps, when viewed as a whole, add nothing significant beyond the sum of their parts taken separately." For these reasons, the Court considered "that the steps are not sufficient to transform unpatentable natural correlations into patentable applications of those regularities." Accordingly, the Court asked whether "the patent claims add enough to their statements of the correlations to allow the processes they describe to qualify as patent-eligible processes that *apply* natural laws?"

The Court concluded that this is not the case, because "the steps in the claimed processes (apart from the natural laws themselves) involve well-understood, routine, conventional activity previously engaged in by researchers in the field." Thus, the Court held the claims invalid for claiming natural laws, which as the Court put it, are not patentable subject matter under 35 U.S.C. 101, thus reviving a long-forgotten position, namely that, due to the fact that 35 U.S.C. 101 mentions only process, machine, manufacture, or composition of matter, laws of nature are implicitly excluded.

In *Association for Molecular Pathology vs. USPTO*,[17] a consortium of plaintiffs challenged couple of patents assigned to Myriad Genetics, namely US 5,747,282, US 5,837,492, US 5,693,473, US 5,709,999, US 5,710,001, US 5,753,441, and US 5,753,441. Next to the USPTO, Myriad acted as a defendant.

The plaintiffs claimed that 15 claims from these seven patents assigned to Myriad were drawn to patent-ineligible subject matter under 35 U.S.C. §101. The District Court of Southern New York revoked the patents in March 29, 2012, stating that they were directed to a law of nature.

In the subsequent appeal proceedings before the Court of Appeals of the Federal Circuit, the Department of Justice (DOJ), interfered and argued for the

---

[16] See footnote 7.

[17] Association for Molecular Pathology v. Myriad Genetics, Inc. No. 10-1406 (Fed. Cir. 2011).

plaintiffs, i.e., against the USPTO. DOJ's federal attorneys suggested what has become notorious as the "magic microscope", according to which (i) a magic microscope can look deep inside cells and find any natural molecule in them, and (ii) any natural molecules that it can find should be excluded from patent protection, since "products of nature" have never been patentable.

Judge Moore, who was a member of the CAFC panel of judges, referred to the magic microscope test as "kitschy". Eventually, in a 2–1 decision the Court revoked the first instance decision, stating, among others, that (1) isolated genes, cDNAs and partial isolated gene sequences are patentable subject matter under §101 as well as (2) methods of screening potential cancer therapeutics by analyzing growth rates of cells with altered BRCA genes in the presence or absence of the treatments. Claims to methods of analyzing BRCA gene sequences and comparing those with cancer-predisposing mutations to normal or wild-type gene sequences were held not to be directed to patentable subject matter.

Not surprisingly, both parties filed a petition seeking rehearing. Eventually, the case was carried to the Supreme Court, who on March 26, 2012 remanded it back to the CAFC for further consideration in light of Mayo Collaborative Services v. Prometheus Laboratories. The latter has, on August 16, 2012, affirmed that isolated human genes are patent-eligible subject matter. The Court emphasized that the Supreme Court's recent decision in Mayo Collaborative Services v. Prometheus Laboratories was not decisive for the instant case, though the Supreme Court's analysis was considered "nonetheless instructive".

However, although they do not relate to stem cells, both cases cast new doubts on the future of hES cell patents, at least in the USA. Notwithstanding this, patent applications related to hEScells are steadily filed, and granted, in the United States

3.3 Non-Patent Related Battlefields

However, patent disputes are only one side of the medal in the USA. Another battle was, and still is, fought namely on the side of public financing. According to the "Dickey Wicker amendment", which was a appropriation bill rider attached to a bill passed by United States Congress during the Clinton administration, the Department of Health and Human Services (HHS) was banned from using public funds for the creation of human embryos for research purposes, or for research in which human embryos were destroyed. Under the Bush administration, the federal financing of research devoted to embryonic stem cells was restricted to 21 already existing cell lines, in order to discourage the use of new embryos for the generation of new stem cell lines. President Obama promised in his electoral campaign to lift these restrictions, in order to expand the number of hES cell lines eligible for federally funded research. In an executive order of March 9, 2009, he instructed the NIH to remove existing limitations on scientific inquiry, and to expand NIH support for the exploration of human stem cell research.

Shortly thereafter, the NIH published draft guidelines allowing funding for research on stem cells derived from donated embryos leftover from fertility treatments. Further, NIH would continue to fund research on adult stem cells and induced pluripotent stem cells. Research on embryos created specifically for research or on stem cells derived by research cloning techniques or by parthenogenesis would not be supported.

A large public discussion followed. In the final guidelines, which took effect July 7, 2009, it was set forth that previously derived stem cell lines that follow the spirit of the new ethical guidelines would be eligible for funding. Further an NIH advisory panel would evaluate these older stem cell lines if needed.

On August 19, 2009, the NIH was sued by two researchers, James L. Sherley, an adult stem cell researcher at the Boston Biomedical Research Institute, and Theresa Deisher, R&D director at AVM Biotechnology in Seattle, before the Federal District Court in Washington.[18]

The claimants, who were backed, among others, by Christian organizations, contended that the funding of embryonic stem cell research would unfairly divert money from adult stem cell research.

As a result, Judge Lamberth issued a preliminary injunction on August 23, 2010, banning federal spending on human embryonic research. The US Government quickly filed an appeal, but in the meantime the NIH had already shut down part of their hES cell research, and stayed grants to researchers that had not yet been paid out. On September 9, 2010 the Appeals Court for the DC Circuit allowed the request to stay the injunction, and the NIH could resume its hES cell programs.

On September 27, 2010, the Appeals Court ruled that the federal funding could go on while the appeals process moved forward, and on April 29, 2011 blocked the decision in a 2–1 ruling, and remanded it back to the District Court. Notably, the dissenter of said decision Judge Lecraft Henderson said her colleagues had performed "linguistic jujitsu". Following this prejudice, Judge Lamberth dismissed the lawsuit on July 27, 2011, thus paving the way for federal funding of hES cell related research. For 2011, the NIH allotted $358 million for non-embryonic stem cell research, and $126 million for embryonic stem cell research.

On August 24, 2012, the Appeals Court for the DC circuit eventually confirmed this decision. Judge Sentelle stated that the Dickey-Wicker act permits federal funding of research projects that utilize already-derived embryonic stem cells, which the court considered are not themselves embryos, because no "human embryo or embryos are destroyed" in such projects.

Thus, the court came at least in one aspect to the same finding as the ECJ and the EBA—namely that embryonic stem cells are not embryos as such.

---

[18] Sherley et al. vs NIH, 1:09-CV-1575.

4 Digressions

4.1 The European Side Battle

As already indicated above, the EPO and the EU are different bodies, and the legal system created under the European Patent Convention (EPC) is generally independent from EU legislation. In order to increase its influence in the patent domain, the EU has however tried to issue directives related to patent matters, which would then become applicable law in the EU member states.

So far, the EU has only managed to issue a directive related to Biotech Inventions (Biopatent Directive 98/44/EG), while an approach to issue a directive related to software patents[19] was dismissed by the European Parliament in 2005.

Through said backdoor, however, the ECJ became the highest instance for issues related to the enforcement of European Biopatents, and for questions related to the validity of national Biopatents, including Biopatents issued through the EPO pathway, too. In all cases the patent-unfriendly attitude and the lack of legal expertise at the ECJ is alarming.

The outcome of the recent Monsanto/Cefetra case,[20] in which the ECJ has dramatically compromised the concept of compound protection, has already confirmed the author's fears related to ECJ's expertise and attitude with respect to patents, particularly to Biopatents (Hüttermann and Storz 2010).

One issue discussed peripherally in the "WARF" case related to the question whether or not the EBA should submit a referral to the European Court of Justice (ECJ) for its opinion on the case. The EBA denied this initiative because (i) neither EPC nor European Law provide a pathway under which EBA can actually send a referral to the ECJ, and (ii) ECJ ruling has no legal effect on the EPO.

On June 15, 2012, a coalition of Patient associations and leading research funders called on the European Parliament to continue EU funding for embryonic stem cell research. The latter is currently debating "Horizon 2020", which is the EU's program for research and innovation for the years 2014–2020. Some provisions in the draft regulation relate to the funding of stem cell research, which is still possible under the current Framework Programme 7. However, these provisions are challenged by delegates who believe that public funds should no longer be spent on embryonic stem cell research.

---

[19] Proposal COM/2002/92/FINAL.

[20] Case C-428/08.

4.2 ECJ and TRIPS

The TRIPS[21] contract is a contract related to the mutual acceptance of IP rights which has been signed by all WTO member states. In TRIPS, the WTO member states have agreed that they will accept particular standards related to patentability, and that patents shall be available for any inventions, whether products or processes, in all fields of technology Art 27 (1). In Art. 27 (2), TRIPS provides the option that member states may exclude from patentability inventions the commercial exploitation of which is may affect protect *ordre public* or morality.

The EU is not a member state of the WTO, but has ratified TRIPS. According to the ECJ, TRIPS has no direct effect on EU legislation. However, the recitals of the Biopatent Directive indicate that TRIPS was taken into account when the directive was drafted.

Should the ECJ exclude hES cells from patentability, this would probably be in line with Art 27 TRIPS, because the *ordre public* issue seems to be a real issue in the member states. However, it is quite unclear if iPS cells and other cells would satisfy this criterion, too.

4.3 When is an Embryo an Embryo?

In his judgement in the Brüstle case, the ECJ decided that "any human ovum after fertilisation, any non-fertilised human ovum into which the cell nucleus from a mature human cell has been transplanted, and any non-fertilised human ovum whose division and further development have been stimulated by parthenogenesis constitute a 'human embryo'" in the meaning of the Biopatent directive;

It is, however, interesting how the ECJ comes to this solution. The approach seems to be a mere teleological approach, because the ECJ does not relate to legal definitions in the different EU member states, but merely constitutes that "Recital 38 in the preamble to the Directive states that the list" of inventions mentioned in the Directive as being contrary to *ordre public* "is not exhaustive, and that all processes the use of which offends against human dignity are also excluded from patentability". Further, the ECJ argues that the "context and aim of the Directive thus show that the European Union legislature intended to exclude any possibility of patentability where respect for human dignity could thereby be affected". The Court thus concludes "that the concept of 'human embryo' within the meaning of Article 6(2)(c) of the Directive must be understood in a wide sense." The ECJ thus defines the term "human embryo" entirely de novo.

EU member states have, however, already defined the term "human embryo" in the past. The German Act for the Protection of Embryos, for example, defines the term under § 8 (1) as a "fertilized, viable human ovum from the beginning of

---

[21] TRIPS is an acronym for "Trade Related Aspects of Intellectual Property Rights".

nuclear fusion, plus totipotent cells extracted from an embryo which can divide under appropriate conditions and develop into an individuum".

Some legal systems see a major cesura 14 days after fertilization. Before this point, the embryo can still be split to develop into two or more children. Further, the embryo has no primitive streak, which is considered as the first step in the development of a central nervous system before that date. This situation has been declared equivalent to a situation in which a patient which has been diagnosed as brain dead and is declared eligible as an organ donor.

This concept is, for example, reflected in the UK Human Fertilisation and Embryology Act 1990, which in section 3 prohibits keeping or using an embryo after the appearance of the primitive streak (Sect. 3a), which is taken to have appeared in an embryo not later than the end of the period of 14 days beginning after fertilization. Dr. Brüstle has used this argument in his ECJ case to define the meaning of the term "embryo" as used in the Biopatent Directive—to no avail, as we all know now.

Interestingly, the Jewish understanding of an embryo which requires utmost protection is completely different. It seems that the Jewish tradition attributes minimal life value to early-stage embryos outside the female uterus. The Talmud defines any embryo up to 40 days old as a mere fluid. Further, it seems also important whether the embryo is inside a woman's uterus or in a lab, where it cannot develop into a child. According to even conservative Rabbis it is thus considered a "*mitzvah*", i.e., a religious mandate, to use those embryos for the benefit of society.

5 Alternatives to hES Cells

In the following, alternatives to hES cells and their potential to avoid the ethical problems raised by hES cells will be discussed. Further, it will be discussed whether these alternatives are, or will be deemed as, patentable subject matter in light of the above decisions. See Table 3 for a summary.

5.1 Cells Obtained by Somatic Cell Nuclear Transfer

This method involves the production of a blastocyst, which was considered as an "embyro" by the EBA and the ECJ. The method further qualifies as a "cloning method", and is thus unpatentable if related to humans.

Before the Brüstle case, no explicit case law existed with respect to these cells, but the BGH referral addressed this issue, too, because the respective cells are mentioned in the Brüstle patent as alternatives to hES cells. As discussed above, the ECJ found this type of cells unpatentable, too. Interestingly, somatic cell nuclear transfer (SCNT) patents granted by the EPO before the ECJ decision relate exclusively to methods or cells referring to non-human animals.[22]

5.2 Stem Cells Obtained by Parthenogenesis

Methods for the production of stem cells by means of parthenogenesis have been seen as a possible solution for the ethical dilemma raised by hES cells. However, the respective methods are still under R&D and not yet ready to be put into practice. The methods involve the production, and destruction, of a blastocyst which is diploid for its maternal genes. For this reason, however, the blastocyst cannot become a viable organism.

While the method is surely not a cloning method, it remained arguable whether or not said blastocyst qualifies as an "embryo", or as a "human body in a stage of formation". The BGH had addressed this issue in his referral, and the ECJ found this type of cells, like cells obtained by somatic cell nuclear transfer.

5.3 Induced Pluripotent Stem Cells

Like stem cells obtained by parthenogenesis, the reprograming of differentiated somatic cells has been seen as a possible solution for the ethical dilemma raised by hES cells. In fact, the reprograming approach avoids the use of human blastocysts, and creates pluripotent cells.

The UK patent office (UKIPO), which is also bound to the provisions of the Biopatent Directive (as the priority date of said application ranks later than the implementation of the Biopatent Directive into UK law)[23] has granted the first iPS-related patent outside of Japan by Jan 12, 2010 to iPierian, which is Bayer Schering affiliate. The patent (GB2450603) relates to an iPS method which involves the use of Klf-4, Oct-4 and Sox-2, but excludes use of c-Myc. The inventor is Kobe-based iPierian researcher Kazuhiro Sakurada.

The outcome of this case has no legal bearing on one parallel EP case (EP2171045), which is still pending, although *ordre public* issues have not been raised in the latter so far. The pending claims in this application require the use of Oct3/4, Sox2 and Klf-4 plus contacting the cells with histone deacetylase inhibitor (HDACi), unlike in the UK, in which the exclusion of c-Myc was sufficient, the applicant preferred to positively recite HDACi, which is discussed as a substitute the oncogene c-Myc in the pertinent literature.

---

[22] According to a study performed by the author.

[23] In the UK, the Biopatent Directive applies to patents filed on or after July 28, 2000, while the priority date of the Sakurada application is June 15, 2007.

Table 2 Overlapping scopes of the EP patents of Shinya Yamanaka and Kazuhiro Sakurada

Yamanaka (Kyoto) EP1970446 B1 (granted)	Sakurada (Kobe) EP2171045 A1 (pending)
Oct and Klf and (Myc and/or cytokine), Sox optional	Oct and Klf and Sox plus HDACi

Interestingly, Shinya Yamanaka of Kyoto University, who pioneered iPS methods, has filed a patent application (EP1970446) which has an earlier filing date than the Sakurada patent. While the main claim as filed recited only Oct, Klf, and Myc, the latest examination report required to also recite Sox, as in a later Yamanaka publication all four factors were considered necessary.

On May 16, 2011, the EPO issued a communication of intention to grant a patent on this application The claims as accepted for grant relate to a nuclear reprograming factor which comprises an Oct gene/gene product, a Klf gene/gene product, plus either a Myc gene/gene product or a cytokine.

The acceptance came quite surprisingly, because initially, the EPO had objected the said claim under Arts 83 and 84 EPC (lack of clarity/lack of enablement), due to a prior publication by the inventor according to which the induction process could be accomplished only by treatment of cells with the four factors Oct, Myc, Klf, and Sox.

However, the applicant has successfully put aside these concerns. Furthermore, moral issues have not been discussed during prosecution, which suggests that, after the Enlarged Board of Appeal decision G2/06 ("WARF", according to which hES cells are banned from patentability as long as their preparation involves the use of a human embryo), the EPO seems to accept that iPS cells do not fall under this exclusion.

Provided the latter is granted with a similar scope, two conflicting patents would exist in Europe, having the following scope: (Table 2).

Groups working under the Sakurada protocol would probably infringe the Yamanaka patent, in case they use a cytokine or Myc in addition to the other factors. In case Myc was left away, and no cytokine was used, the Yamanaka patent would probably not be infringed.

On the other hand, groups working under the Yamanaka protocol would probably infringe the Sakurada patent, if granted as shown above, in the event that they use Sox and HDACi in addition to the other factors.

Companies working with iPS cells should be aware of this confusing situation and ask for expert counsel before they enter the marketplace with their products.

In Feb 2011, iPierian announced that they have entered into a series of IP agreements with Kyoto University, home of iPS pioneer Shinya Yamanaka, under which iPierian assigns its iPS portfolio to Kyoto University, while the latter grants non-exclusive worldwide rights to its iPS portfolio for use in drug discovery and development. As part of the agreement, Professor Shinya Yamanaka has joined iPierian's Scientific Advisory Board.

This deal illustrates the complicated relationship between both patent portfolios. Further, the deal demonstrates that, at least today, the major goal of iPS technologies is drug development (rather than the oft-cited regenerative medicine).

On May 3, 2012, an opposition was filed against Kyoto University's Yamanaka patent EP1970446. The opposition was lodged by UK law firm Olswang LLLP, without disclosing the true claimant. Proceedings are ongoing at the editorial date.

However, the EPO does not seem to consider iPS cells as falling under the exemptions related to human embryos, and their use. The ECJ has not discussed this type of cells in his the Brüstle decision, and it can thus be considered as granted that these cells remain patentable in Europe, being subject to the same bars as inventions from other fields of technology, i.e., novelty, inventive step, and industrial applicability.

6 Adult Stem Cells

Adult stem cells, also called "Pluripotent germline stem cells (pGSs)", are pluripotent cells which will, without further steps, not redifferentiate to totipotent cells.

Examples are spermatogonial Stem cells or pluripotent somatic Stem cells, e.g., from umbilical chord, from skin of people with Friedreich's ataxia, or enterogastric neural stem cells. It is unlikely that ECJ will consider these cells as "embryos", or as a "human body in a stage of formation". The BGH has not addressed this issue, so the ECJ will most probably not opine on these cells.

7 Cells Obtained by Transdifferentiation

Vierbuchen et al. (Nature 463, 1035-1041) have transdifferentiated fibroblasts to nerve cells with only two transcription factors (BAM and BAZ). This means that no detour via pluripotent cells, stem cells, or blastocysts is necessary. The cells do thus never reach a stage which can be considered "embryonic" according to common understanding.

It is, thus, quite unlikely that ECJ will consider these cells as "embryos", or as a "human body in a stage of formation". However, the BGH has not addressed this issue in his referral and, accordingly, the ECJ has noted considered these cells in his decision.

8 Methods Which Allow the Production of hES Cells Without Destroying an Embryo

What if methods were available which allow the production of hES cells without destroying an embryo? In a letter to former President Bush, Leon Kass, who was the chair of the President's Council on Bioethics, envisaged in May 2005 that such

Table 3 Different methods to derive stem cells, or to transdifferentiate cells. Shaded areas sow what the ECJ considers as an embryo

Type	Potency status	Human embryonic stem cells	Cells obtained by parthenogenesis	Cells obtained with SNCT	Adult stem cells	Induced pluripotent stem cells	Cells obtained by transdifferentiation
Oocyte	Totipotent						
Blastomere and/or blastocyst	Totipotent						
Stem cell	Pluripotent						
Progenitor cell	Multi-oder oligopotent						

approach would probably guide a way to a solution out of the ethical dilemma posed by hES cell research.

Later in 2005, S. Matthew Liao discussed for the first time the "Blastocyst Transfer Method". He hypothesized that this method, in which cells from the inner cells mass of a blastocyst (<125 cells) could be extracted and used for the production of hES cells without destroying the latter and, specifically, without harming its chance of developing into a healthy functioning individual.

In 2006, Klimanskaya et al. reported about the successful derivation of hES cells from cells obtained by biopsy of a human blastomere (8–20 cells), which survived this incident. This approach has been termed "Blastomere extraction".

Both authors discussed the possibility that the embryo could still be implanted and brought to term. Notably, a similar approach is already applied in preimplantation genetic diagnosis (PGD), where one or more cells are sampled from a blastocyst obtained by in vitro fertilization and undergo molecular screening, while the remaining blastocyst is then implanted into a mother. Success rates of 44 % have been reported for such approach, e.g., by the Guy's and St Thomas' Centre for Preimplantation Genetic Diagnosis.

Although PGD, and the invasive treatment the embryo is subjected to therewith, finds increasing acceptance among parents, it is hard to imagine that the latter would agree with such treatment only to allow researches to obtain hES cells from their embryo.

This means that, although an embryo would probably survive such treatment, it is unlikely that it would be implanted thereafter. It would rather have to go back into the freezer to preserve, at least theoretically, its potential to develop and differentiate to a fetus, and eventually be born. Even then, however, one would probably assume that the embryo has not been killed or destroyed by the extraction process.

Although different authors object these approaches as like ethically problematic (Holm 2005) and thus unsuitable to render hES cell research ethically acceptable, these approaches could at least bypass the exclusion set forth by the Biopatent directive under Art 6(c), according to which the use of human embryos for industrial or commercial purposes shall be considered unpatentable. While embryos would still be "used" in such process, they would at least not be "destroyed", how the ECJ has put it in the Brüstle decision.

However, the ECJ has also ruled that inventions using human embryos are patentable for therapeutic or diagnostic purposes which are applied to the human embryo and are useful to it. However, even if the suggested methods of Blastocyst Transfer and Blastomere Extraction do not destroy the embryo, they are are most probably not useful to the embryo itself.

As the ECJ ruled that the exclusion from patentability also covers the use of human embryos for purposes of scientific research, (because the grant of a patent imply its later industrial or commercial application) it is quite likely that obtaining hES cells with Blastocyst Transfer and Blastomere Extraction would still be considered exempt from patent protection by the ECJ Table 3.

References

Bergman K, Graff GD (2007) Nat Biotechnol 25:419–424
Caulfield T et al (2009) Nature reports stem cells doi:10.1038/stemcells.2009.61
Holm S (2005) Am J Bioeth 5(6):20–21
Hüttermann A, Storz U (2010) Monsanto soy bean patent cases—a paradigm shift gathering in case the ECJ takes over patent jurisdiction, Les Nouvelles, pp 156–159
Kalorma Information (2012) Stem cells: worldwide markets for transplantation and cord blood banking, New York
Klimanskaya I et al (2006) Nature 444:481–485
Liao SM (2005) Rescuing human embryonic stem cell research: the blastocyst transfer method. Am J Bioeth 5(6):8–16
Rabin S (2005) Nat Biotechnol 23(7):817–819
Visiongain Ltd (2012) Stem cell technologies: world market outlook 2012–2022, London

The Limits of Patentability: Genes and Nucleic Acids

Aloys Hüttermann

Abstract The patentability of genes and nucleic acids in the EU is ruled by the "Biotech directive" 98/44/EG, which allows the patenting of encoding genes and nucleic acids under certain requirements, the main being that the function of the protein for which the gene encodes is known. Non-encoding genes and nucleic acids are treated like usual chemical substances. The protection for genes has been limited to purpose-bound protection only by the "Monsanto"-Ruling of the ECJ.

Keywords Biotech · IP · Patents · Biotech directive · ECJ

1 Introduction

When discussing the patentability of Genes and Nucleic acids, the following problems and thoughts immediately arise, which make the patentability a quite complicated issue.

(a) Although Nucleic acids are definitely chemical substances, there is nearly no need for patenting Nucleic acids *as such* although some Nucleic Acids have e.g., been used as aptamers or as biocatalysts in science.[1] Nucleic Acids are

---

[1] cf as examples Burmeister et al. (1997) or Eckardt et al. (2002).

A. Hüttermann (✉)
Michalski Huettermann and Partner Patent Attorneys, Neuer Zollhof 2,
40221 Duesseldorf, Germany
e-mail: ah@mhpatent.de

A. Hübel et al., *Limits of Patentability*, SpringerBriefs in Biotech Patents,
DOI: 10.1007/978-3-642-32841-1_3, © The Author(s) 2013

interesting primarily because they encode information (e.g., for a protein)—the *information* is what raises the interest, not the chemical substance "Nucleic Acid".[2]

(b) The most interesting Nucleic acids are not surprisingly genes which occur in nature like human genes. However, this raises the question in this context how to distinguish between inventions (which are patentable) and discoveries (which are not).

(c) The patenting of Genes and Nucleic acid is a field where political interests and ethical topics are more prominent than in most other technical fields as e.g., the patenting of new chemical substances for organic light diodes. This has led to increased pressure to push the ability and possibility as well as the scope of patenting Genes and Nucleic acids in the desired direction by many pressure groups and lobbies.

Therefore, the patentability of Genes and Nucleic acids can be considered somewhat of a special field in patenting where the usual rules cannot be applied so easily.

2 The Biotech Directive

The patenting of genes and Nucleic Acids in Europe is ruled primarily by the so-called "Biotech directive" 98/44/EG dated 6 July 1998, which was adapted into German law only in 2005. All member states of the EU have implemented this directive into their national laws.

The existence of a directive has the effect that the highest juridical body in the field of biotechnology is the ECJ; this topic deserves some more thorough discussion:

The ECJ becomes competent on the basis of EU-regulations and EU-directives. EU-regulations are "direct" law, i.e., they have a uniform wording for the whole EU and become law directly. Examples are the Regulations concerning the Community Trademark[3] and Community design.[4]

EU-directives are "indirect" law, i.e., each EU member state must implement it into its national law; however, usually the directive at certain points leave it to the member state how they want to regulate it, so that there is some "play" for the member states. Furthermore, it usually takes some years until all member state have implemented the directive, a few are even never implemented at all. Examples for such directives are the Design directive[5] and the Trademark direc-

---

[2] This fact shows interesting parallels between Biotech and Software patenting, cf. Hüttermann and Storz (2009).

[3] The latest is Council Regulation EC 207/2009 February 26th 2009 (the original is EC 40/94).

[4] Council Regulation (EC) 6/2002 December 12th, 2001.

[5] Directive 98/71/EG, October 13th 1998.

tive,[6] which leads to the result that in the EU for both Design and Trademark matters—regardless whether national or community—the ECJ has the final say.

Although not a body of the European Union, the EPO also uses the Biotech directive and the Rules 26–34[7] are a wordical copy. It was implemented already on September 1, 1999 and is used ever since. However, the EPO has in a landmark decision[8] refused to ask the ECJ for instructions when deciding about the patentability of biotechnological patent applications.

Therefore at this moment, the ECJ has the final say on the Biotech sector for national patent applications and for infringement matters of national patent applications *and* European patents, if the litigation takes place in an EU member state—but *not* in the granting process of European Patents before the EPO.

The Directive deals with all sorts of biotechnological inventions and cannot be discussed wholly, however the following should be noted:

In the directive it is decided that genes *as such* cannot be patented, however *isolated* genes can, even if they are identical with a gene which occurs in nature. What does "isolated" in this context mean? This is usually interpreted as that a gene has been "cut-out" out of a chromosome and it has been identified which parts of the chromosome form a specific gene and which neighboring regions e.g., belong to another gene or are non-coding.

Furthermore, to be patentable, genes must be industrial applicable. This is a requirement which is—although demanded for every invention to be patentable—seldom an issue because usually nearly every apparatus is somehow industrially applicable. In the context of gene patenting, however, industrially applicable means especially that the function of the protein that the gene encodes for is known.

In the biotech directive it is, furthermore, demanded that this function must be specified in the application which more or less means that the function of the protein must be known at the time of filing and cannot be added later. This is also the wording of Rule 29 EPC.

However, *inter alia* the German government decided in its implementation of the Biotech directive[9] that the function must be inserted in the claim, i.e., that substance protection for a given nucleic acid is not possible, only a use protection.

Since the EPO does not demand such a limitation it is not surprising that the number of national German applications for genes or nucleic acids is and has been close to zero with the applicants filing at the EPO instead.

---

[6] Directive 89/104/EWG, December 21st 1988.

[7] former Rules 23b to e, 27a and 28/28a EPC.

[8] G 02/06, of 25 November 2008, OJ EPO 5/2009, 306 332. The patentability of the parallel German patent, however, was finally decided by the ECJ.

[9] §1a of the German Patent law. For a more detailed analysis cf e.g., Hüttermann and Storz (2005).

3 Practice of Prosecution of Gene/Nucleic Acid Application

The EPO's current practice on patenting of gene sequences or Nucleic acids can be outlined approximately as follows:

3.1 Formal Requirements/Sequence Listing

The EPO requires for any patent application which contains a sequence that a sequence listing is filed with the application, otherwise a fee for late filing applies or the application is withdrawn. This goes also for applications where the sequence is not part of the claims.

3.2 Patenting of Encoding Gene Sequences "As Such"

The EPO allows substance claims on encoding gene sequences. However, if the gene encodes for a protein, the function of the protein must be known and specified in the application. A simple statement that the protein must have some function has been not accepted by the EPO e.g., in the "New Seven-Transmembrane-Receptor V28"-decision[10]

For novelty of the gene sequence, it is sufficient that the encoded protein is novel, even if the sequence as such has been known (e.g., in that it was disclosed in the HUGO database or Lawn Library). Inventive step can be acknowledged if the protein was different to isolate or has surprising properties.

The EPO allows not only claims on the gene sequence as such but also on "derived" gene sequences. A whole cascade might look like the following:

1. A nucleic acid molecule, selected from the group consisting of

   (a) a nucleic acid molecule comprising a **nucleotide sequence** presented as SEQ ID NO: 1
   (b) a nucleic acid molecule **encoding a polypeptide comprising the amino acid sequence** presented as SEQ ID NO: 2, wherein said polypeptide is a.../has a...activity
   c) a nucleic acid molecule that is a **fraction, variant, homolog, or derivative** of the nucleic acid molecules of (a-b)
   (d) a nucleic acid molecule that is a **complement** to any of the nucleic acid molecules of (a-c)

---

[10] Decision of the Opposition Board of 20 June 2001, c.f. OJ EPO 2002, 6, p. 293–308.

(e) a nucleic acid molecule that is **capable of hybridizing** to any of the nucleic acid molecules of (a-d) under stringent conditions

(f) a nucleic acid molecule which comprises, in comparison to any of the nucleic acid molecules of (a-e) at least one **silent single nucleotide substitution,**

(g) a nucleic acid molecule according to (a-f) which is **code optimized** for a given expression host,

(h) a nucleic acid containing the encoding cDNA insert of the plasmid contained in **ATCC XXXXX** and/or

(i) a nucleic acid molecule having a **sequence identity of at least 70, preferably 95 %** with any of the nucleic acid molecules of (a-h).

The last point (i)—the protection of non-identical but similar gene sequences—has been discussed thoroughly and the level/percentage of identity has been oscillating with the EPO allowing claims on 50 % identity or demanding a claimed percentage of 90 %. In the last years, however, a 70 % identity level seems to be acceptable in most constellations unless extraordinary circumstances apply. However, so far there is no case law if such a non-identical gene sequence is really a patent infringement.

3.3 Patenting of "Derived" Substances

Furthermore if the patentability of the gene sequence has been established, the EPO usually allows not only "derived" gene sequences but also "derived" other substances such as:

2. A **vector** comprising the nucleic acid according to claim 1.
3. A **host cell** transformed with the vector of claim 2 or the nucleic acid of claim 1.
4. A **method for producing a protein**, comprising the steps of culturing a cell according to claim 3 under conditions which allow expression of a protein.
5. A **protein** selected from the group consisting of

   (a) a protein encoded by a nucleic acid according to claim 1,

   (b) a protein comprising the amino acid sequence according to claim 1,

   (c) a protein according to (a) or (b) which comprises at least one conservative amino acid substitution,

   (d) a protein obtainable by the method of claim 4.

6. An **Antibody** specifically recognizing the protein of claim 5, or a fragment or derivative thereof.

The EPO considers claims as these to be patentable because the production of (e.g.) a protein if the gene sequence is known, or an antibody if the protein is known, is considered standard practice and involves no inventive step from any skilled person in the art.

3.4 Patenting of Non-Encoding Gene Sequences

There exist fewer patent applications and patents on non-encoding gene sequences. However, non-encoding gene sequences (such as primers, Antisense-DNA, promoters, miRNA) can be subject of a patent as well. For these substances, the usual rules as for chemical compounds apply.

4 Litigation with Respect to Nucleic Acid Patents

Considering the fact that the PCR-technique which has led to a new biotechnological revolution was developed in the 1980s and that the juridical background is from the late 1990s it is not surprising that there are just a few cases which deal with the special conditions and characteristics of gene patents since that requires both a valid patent (which requires an invention) and an alleged infringement.

However, there exist some case law which sheds a light on the problematics concerned with gene patents and which is discussed in the following.

4.1 The "Monsanto" Case

This is the first ruling of the ECJ on gene patents and is, therefore, the most important case law which exists so far. It is valid Europe-wide. For this reason, this case deserves a more detailed discussion.

4.1.1 Case Background

The EP 546090 B1 on which the infringement litigation was based protects a DNA sequence which was implemented in soy beans.[11] The respective case is related to

---

[11] The following is an abridgement of the case which has some more complicated aspects to it. For further information cf. Hüttermann and Storz (2010) and Kock (2010).

the import of soy bean meal from Argentina to the Netherlands. The meal has been produced form genetically modified soy beans originally provided by Monsanto, and grown in Argentina. These soy beans carry the DNA sequence described in the EP 546 090 B1[12] which provides a resistance against Monsanto's Roundup herbicide so that the herbicide can be used together with the soy beans. While Monsanto has no patents protecting the said soy beans in Argentina, the cited patent for the DNA is in force in Europe.

Monsanto had the border police stop a ship with imported soy meal from Argentina in a Dutch port and tried to sue the importer, Cefetra BV, for patent infringement in the Netherlands, among others, on the basis of the following claim:

    An isolated DNA sequence encoding a Class II EPSPS enzyme selected
    from the group consisting of SEQ ID NO: 3 and SEQ ID NO: 5.

Monsanto held that this claim is infringed by the imported soy bean meal as the respective DNA sequences can be found in the meal, at least in traces. The Dutch court then referred to the ECJ as it found that the respective legal issues require an interpretation of the Biopatent Directive[13] and asked the ECJ (inter alia) the following question:

    Must [the directive] ... be interpreted as meaning that the pro-
    tection provided under that provision can be invoked even in a
    situation such as that in the present proceedings, in which the
    product (the DNA sequence) forms part of a material imported into
    the European Union (soy meal) and does not perform its function at
    the time of the alleged infringement, but has indeed performed its
    function (in the soy plant) [...]?

4.1.2 The Case Before the ECJ

Monsanto's position in the court procedure before the ECJ may be (very roughly and abridged) summarized as:

- Although the patented DNA is present in the meal only as undesired impurity, it is there and therefore the patent is infringed, since the patent claim as such allows substance protection. The function of the DNA has been clearly identified in the patent specification (as demanded by the Biopatent directive) and undoubtedly during the breeding of the soy beans it has already performed it. So, the question about the functionality of the protected DNA is not an issue.
- It may be discussed whether the Biopatent directive as such is not even needed since DNA is also a chemical compound, for which absolute substance protection is available.

---

[12] The protected DNA encodes for an enzyme called "5 enolpyruvylshikimate 3 phosphate synthase".
[13] Monsanto vs. Cefetra et al., C-428/08. Parallel cases were pending in the UK, Denmark, and Spain.

- Since there is no *de minimis* provision in the Biopatent directive, the fact that the DNA is only present in traces is of no importance.

In his plea, the Attorney General did not follow this position, actually he stated that in the view of the Biopatent Directive substance protection for DNA may not be available at all. Since it is correct that there is no *de minimis* provision, he referred to the function requirement, as it is recited in Art. 9 of the Directive, according to which

> the protection conferred by a patent on a product containing or consisting of genetic information shall extend to all material [...] in which the product [is] incorporated and in which the genetic information is contained and performs its function.

Since the DNA does not perform its function in the soy meal, there would be no patent infringement.

Although Monsanto tried to avoid a final verdict of the ECJ by withdrawing the case shortly after the Attorney General's statement was published the ECJ nevertheless issued a judgment on July 6, 2010. In this judgement, the Attorney General's opinion in view of the denial of an infringement was confirmed, however the ECJ even shortcut the Attorney General's argumentation by stating that:

> Since the Directive thus makes the patentability of a DNA sequence subject to indication of the function it performs, it must be regarded as not according any protection to a patented DNA sequence which is not able to perform the specific function for which it was patented [...] An interpretation to the effect that, under the Directive, a patented DNA sequence could enjoy absolute protection as such, irrespective of whether or not the sequence was performing its function, would deprive that provision of its effectiveness. Protection accorded formally to the DNA sequence as such would necessarily in fact extend to the material of which it formed a part, as long as that situation continued.[14]

and therefore concluded that the answer to the above question is "no". The result is in fact the introduction of purpose-bound protection for DNA sequences "in through the back door",[15] i.e., at the time being absolute substance protection for DNA can be considered a thing of the past.

It is in my personal opinion a sad fact that neither the Attorney General nor the ECJ seem to have thoroughly studied the available case law on chemical substances, because by doing so they could have solved the situation and the complex of problems associated with the fact that the DNA was only present as impurity to the obviously desired result without giving up the concept of absolute substance protection. There is

---

[14] ECJ, Judgement of 6 July 2010 in the case C 428/08, [47–49].

[15] to quote Michael A. Kock, cited above.

a ruling by the Düsseldorf Court which could have served as a reference case[16] but was never mentioned neither in the plea nor in the judgement at all.

4.1.3 A short Excursion: The Parallel Case in the UK

Just as a side remark it should be noted that there was a parallel case in the UK. Although the patent in suit as well as the case background was nearly analogous, the case was not referred to the ECJ. Rather, the judge on trial Judge Pumfrey denied infringement due to the lack of the feature "isolated" in the Claims (which as discussed above is a term arising out of the Biotech directive). Since the term "isolated" was not defined in the patent specification, he concluded that:

> 77. [...] in the claims the word ``isolated'' has precisely the meaning [...] ``separated from other molecular species in the form of a purified DNA fragment''.

Of course, if this feature is interpreted like that then in the soy meal there cannot be an infringement because the DNA is "mixed" with the meal and not separated. As a consequence, at least to my knowledge I have strong problems to imagine any situation where an infringement would be affirmed since DNA is—even in laboratories—usually never "isolated" but present in aqueous solution, mostly together with buffers and DNAse inhibitors to prevent its degradation.

Since "isolated" is a commonly used term in DNA patents, this may be even a worse ruling for patent owners that the verdict of the ECJ.

4.1.4 Result

As stated above, the result of the case is that although the EPO grants "absolute" patents on gene sequences, these patents are unenforceable to that extent that they go beyond purpose bound protection.

4.2 The "Amgen" Case

This case[17] was handled before the UK and has de jure validity only for the UK. However, since this was one of the first cases concerning infringement in the field of biotechnology, it has gained some amount of attention Europe-wide. This case shows very well the problems related to patenting of gene sequences as well as proteins related to these gene sequences and the constellations which quickly arise.

The patent in suit was related to erythropoietin (EPO), which is a protein involved in the production of red blood cells. EPO has been on the top-ten list of the most important pharmaceuticals for quite a number of years.

---

[16] LG Düsseldorf "Grasherbizid", GRUR 1987, 896, cf. Hüttermann and Storz (2011).

[17] For a detailed discussion cf. Brandi-Dorn (2005).

Kirin-Amgen, a biotech company, had patented sequences relating to EPO in a patent. However, the properties of EPO and EPO as such had been known prior to the filing date of the patent; only a gene encoding for EPO was not disclosed. After finding such a gene in 1983, a patent was filed which was granted by the EPO in 1998 (!).

Immediately after issuance of the patent, Kirin-Amgen sued TKT, another biotech company, for patent infringement.

The patent had three independent claims, which have to be considered equally and independently:

A *first* claim was directed to several *DNA sequences* as such (which were all exogenic).

A *second* claim was directed to an *EPO-polypeptide* having certain properties, especially having a higher molecular weight than EPO-polypeptide isolated out of urine.

A *third* claim was directed to a *polypeptide* product of an expression of any of the DNA-sequences of the first claim in a eukaryotic host cell.

The problem in the infringement suit was the different production method of the EPO protein.

Amgen produced EPO by exprimation of a modified eukaryotic host cell using one or more of the patented DNA sequences.

TKT produced EPO by first producing cells which had been modified with a promoter, which induced the production of endogenic EPO, then producing EPO using the cells and finally importing the EPO *inter alia* into the UK.

2004, the courts finally ruled that TKT would not infringe any of the claims of Amgen-Kirin. The given reason is different for all of the three claims involved:

Concerning the first sequence claim the court ruled that TKT would not make use of any of the (exogenic) DNA sequences. The use of a promoter would produce EPO by using the endogenic DNA, which was not subject of the patented claim. The fact that the genome as such must be known in order to make the promoter, and therefore somewhat indirect TKT was using the technical knowledge of Amgen-Kirin was not considered a patent infringement.

Concerning the second claim the court ruled that this claim was not supported by the description and technical teaching of the patent. Endogenic EPO could not be patented since it was known before the filing date. Amgen-Kirin had therefore differentiated the claim by introducing that "their" EPO-polypeptides should have a higher molecular weight than the isolated EPO. This simple differentiation was not considered to be enough to specify the EPO-polypeptide for which protection was sought, especially after it was found out that most EPO-polypeptide exprimed by the protected DNAs have a lower molecular weight than endogenic EPO. For this reason, the second claim was held invalid, thus there was no patent infringement of this claim

Concerning the third claim the court ruled that this claim was invalid as well, since it was not ruled out that any exprimated polypeptide using one of the protected DNA sequences would not be identical with the (known) endogenic EPO. Even if the DNA sequences are all different, this does not necessarily mean that the product of the exprimation is different from the prior art. For this reason, finally also the third claim was held invalid, thus there was no patent infringement of this claim.

As a result, the first claim was not infringed and the other two claims were held invalid—which results in an overall non-infringement of the patent.

4.3 The "HGS/Eli Lilly" Case

This case was handled before the UK and has de jure validity only for the UK. However, since this was the first case where the question of broad gene patents was handled outside the EPO, it has gained some amount of attention Europe-wide, too.

The patent in suit was an European Patent directed to inter alia the nucleotide sequence for a protein (now) called Neutrokine-α[18] which was believed to be involved *inter alia* in inflammatory diseases and other immune responses. Patentee is Humane Genome Sciences ("HGS") Ltd., a company which uses bioinformatics to derive e.g., genetic information out of a known gene sequence and which has filed several hundred patent applications on gene sequences. At the time of filing, the true properties of Neutrokine-α had not been known, only anticipated from bioinformatics (and some tissue distribution experiments) which suggested that Neutrokine-α belongs to the so-called "TNF ligand superfamily" of which the desired properties (e.g., involvement in inflammatory diseases) are known. No in vitro or in vivo data were included in the application nor the patent, still the EPO had granted a patent on the sequence.

Neutrokine-α has kept its promises and is now one of the most interesting targets in pharmaceutics. The pharmaceutical company Eli Lilly, interested in developing an antibody against Neutrokine-α[19] filed an invalidation in the UK against the patent, arguing that the patent was invalid due to the lack of data about Neutrokine-α at the time of filing.

Both in first and second instance Eli Lilly's view was affirmed and the patent declared invalid, *inter alia* for lack of industrial applicability (as described above, industrial applicability is denied when no function of the protein is known for which the gene sequence encodes).

However, before the UK Supreme court the ruling was reversed and the patent was upheld as granted. It was ruled that the information given in the patent was enough to fulfill all requirements for patentability.

---

[18] At the time of filing this protein had no real name, it was named Neutrokine α only years after the filing date.

[19] As described above, the EPO routinely grants antibody claims in conjunction with gene sequence claims; this was also the case here and the reason why Eli Lilly started the court action.

5 Summary

The practice and rules on the patenting of Genes and Nucleic Acids are different than on other chemical substances as far as encoding gene sequences are involved, due to the special nature of these compounds. Besides ethical issues problems arise out of the fact that it is usually the code which is interesting, not the substance as such.

In Europe, the patenting of gene sequences is ruled primarily by the so-called "Biotech directive" 98/44/EG, which de jure allows the patenting of gene sequences for isolated Nucleic acids if the function of the encoded protein is known. The EPO routinely also grants claims on "derived" gene sequences and products such as the encoded protein or an antibody against it. The function of the protein must be disclosed in the application.

However, the scope of protection of gene sequences has been limited to purpose-bound protection by the ECJ's ruling in the "Monsanto"-Case.

For non-encoding Nucleic acids one can generally say that the usual rules as for other chemical compounds apply.

References

Brandi-Dorn M (2005) Der Schutzbereich nach deutschem und britischem Recht; die Schneidmesser-Entscheidung des BGH und die Amgen-Entscheidung des House of Lords. Mitt, pp 337–343

Burmeister J, von Kiedrowski G, Ellington A (1997) Cofactor-assisted self-cleavage in DNA libraries with a 3'-5'-phosphoramidate bond. Angew Chem Int Ed Engl 36:1321–1324

Eckardt LH, Naumann K, Pankau WM, Rein M, Schweitzer M, Windhab N, von Kiedrowski G (2002) DNA nanotechnology: chemical copying of connectivity Nature 420(6913):286

Hüttermann A, Storz U (2005) Zu den Auswirkungen der Änderungen des Patentgesetzes durch die Umsetzung der Biopatentrichtlinie. Zeitschr f Biopolitik 45–52

Hüttermann A, Storz U (2009) A comparison between biotech- and software-related patents. Eur Intellect Prop Rev 31(12):589–592

Hüttermann A, Storz U (2011) Die möglichen Auswirkungen des Monsanto-Urteils des EuGH auf das Konzept des Stoffschutzes bei chemischen Verbindungen. Mitt d dt Patentanwälte 1:1–4

Hüttermann A, Storz U (2010) Monsanto soy bean patent cases—a paradigm shift gathering in case the ECJ takes over patent jurisdiction. Les Nouvelles 4(9):156–159

Kock M (2010) Purpose-bound protection for DNA sequences: in through the back door? J Intellect Law Pract 7:495–513

About the Authors

Dr. Ulrich Storz was born in 1969 in Muenster. He graduated in Biology at the University of Muenster in 1998, where he received his Ph.D. in 2002. He is author and co-author of several scientific publications in the field of biology and biophysics as well as of several juridical publications in the field of intellectual property. He passed the German Patent Bar Examination in 2005. Since 2005, he has been admitted to practice as European Trademark Attorney at the European Trademark Office (OHIM). In 2006, he has been registered in the list of representatives before the European Patent Office. Main practice areas in the field of Intellectual Property Law include Patent Prosecution, FTO and Patent Infringement, as well as Patent strategies; especially in the Life Science field (i.e. Biotechnology, Biophysics, Biochemistry, Microbiology). One of his major fields of interest is Antibody IP.

Ulrich Storz is active as a speaker for the congress management company "Forum Institut für Management GmbH", and he organizes the annual Rhineland Biopatent Forum.

Dr. Andreas Hübel born in 1964, studied biology at the University of Hamburg. After graduation, he worked at the Bernhard-Nocht-Institute for Tropical Medicine and received his Ph.D. in 1996 for a thesis in the field of molecular microbiology. He worked as postdoctoral fellow at Harvard Medical School and Washington University School of Medicine before joining the Center for Hygiene at Philipps-University Marburg. He is co-author of several scientific articles in the fields of molecular neurobiology, molecular parasitology, and glycobiology.

A. Hübel et al., *Limits of Patentability*, SpringerBriefs in Biotech Patents, DOI: 10.1007/978-3-642-32841-1, © The Author(s) 2013

Andreas Hübel began his professional career in intellectual property in 2000 and was admitted to bar as German Patent Attorney and as European Patent Attorney in 2005. As chartered European Trademark and Design Attorney he is also admitted to represent clients before the Office of Harmonization for the Internal Market.

The main fields of his activity in intellectual property comprises drafting and prosecuting patent applications, attacking, or defending patents, and assessing validity of intellectual property rights and Freedom-to-Operate in the areas of pharmaceuticals, biochemistry, biotechnology, and adhesives.

Dr. Aloys Hüttermann born in 1972, studied Chemistry at the University of Freiburg and received his diploma degree in 1997. He finalized his Ph.D. thesis in 2001 related to synthetic organic chemistry.

Aloys Hüttermann is co-author of the textbook "Das Basiswissen der Organischen Chemie" (the basic knowledge of organic Chemistry, published at Wiley-VCH) and of several legal publications in the field of intellectual property.

He passed the German Patent Bar Examination in 2005. Since 2005, he has been admitted to practice as European Trademark Attorney at the European Trademark Office (OHIM), since 2006 he is European Patent Attorney.

Aloys Hüttermann is a founding member and partner of Michalski Hüttermann & Partner Patent Attorneys.